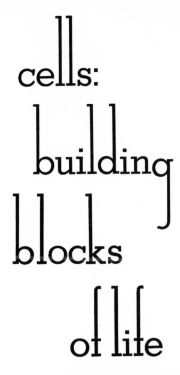

cells: building blocks of life

by Dr. Alvin Silverstein and
Virginia B. Silverstein

illustrated by
George Bakacs

PRENTICE-HALL, INC.
Englewood Cliffs, N.J.

For Gertrude and Samuel Opshelor

Fourth printing.........September, 1971

CELLS:BUILDING BLOCKS OF LIFE
by Dr. Alvin Silverstein and Virginia B. Silverstein

13-121715-1

Library of Congress Catalog Card Number: 69-14810

Printed in the United States of America • *J*

Prentice-Hall International, Inc., London
Prentice-Hall of Australia, Pty. Ltd., Sydney
Prentice-Hall of Canada, Ltd., Toronto
Prentice-Hall of India Private Ltd., New Delhi
Prentice-Hall of Japan, Inc., Tokyo

CONTENTS

Also by the authors:

The Respiratory System:
  How Living Creatures Breathe
The Circulatory System:
  The Rivers Within
The Digestive System:
  How Living Creatures Use Food
The Nervous System: The Inner Networks
The Sense Organs:
  Our Link With the World
The Endocrine System: Hormones in the
  Living World
The Reproductive System:
  How Living Creatures Multiply
Life in the Universe
Rats and Mice: Friends and Foes of Man
Unusual Partners:
  Symbiosis in the Living World
The Origin of Life
A Star in the Sea
A World in a Drop of Water
Carl Linnaeus: The Man Who Put
  the World of Life in Order
Frederick Sanger: The Man Who
  Mapped Out a Chemical of Life
Germfree Life:
  A New Field in Biological Research
Living Lights:
  The Mystery of Bioluminescence
Bionics:
  Man Copies Nature's Machines
Harold Urey: The Man Who
  Explored from Earth to Moon
Metamorphosis: The Magic Change
Mammals of the Sea

In preparation:
  The Muscular System:
    How Living Creatures Move

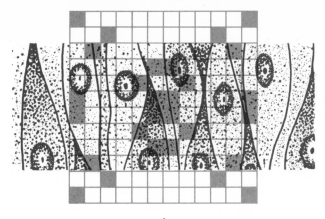

1
What Is a Cell?

Look about you. The people in the street, the birds sitting on the branch of a tree, and even the tree itself—all of these have something in common. They and all other living things are made up of tiny building blocks of life, called *cells*.

These cells are so small that you need a microscope to see them. And yet they are much more complicated than wooden building blocks.

Just as blocks can be used to build various structures—houses and bridges and tunnels—nature uses her building blocks, the cells, to make more than a million different kinds of living creatures. From tiny cells are made huge whales, larger than a house. But the cells in a timid little mouse, hiding in a dark corner, are just as large as the cells in the whale. For it is not the size of an

1

animal's cells that makes it large or small, but how many cells it has.

Plants, too, are made of the same kind of building blocks. The giant redwood trees in California are made of tiny cells, and so is the grass on the front lawn.

But what *is* a cell? Just as the words of our language are made up of letters, our bodies are made up of cells, the basic units of life. And just as there are different letters in the alphabet, there are many kinds of cells. They have different sizes and shapes, and they have different functions.

Some cells look like tiny ice cubes; others are long and thin, and may stretch for inches or even a few feet. There are cells that look like little rods or balls, or doughnuts without the hole in the middle. Others look like kites or tadpoles. Some have no fixed shape at all, but merely look like blobs of jelly.

Some cells carry messages to and fro in our bodies. Others help us to walk and talk by pulling on our bones. Still others fight germs and help to keep us well.

Although most cells are smaller than the point of a pin, each is a busy factory. There are thousands of different chemicals in a cell, and each one seems to have a special job to do. Something is happening

2

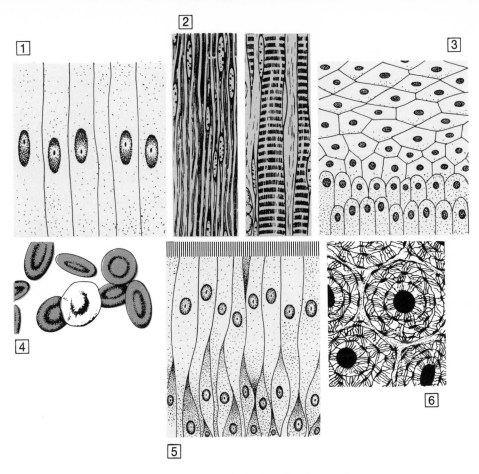

Different kinds of cells: 1-epithelial cells from the stomach lining; 2-muscle cells from the stomach wall and
from a muscle that pulls on a bone; 3-epidermal cells
from the skin; 4-red blood cells and one white blood cell;
5-epithelial cells from the lining of the windpipe; 6-bone-
cells.

all the time. Chemicals are being freshly made,
and other chemicals are passing in or out of the cell
through a special wall called the *cell membrane.*

3

Just as a factory has a main office, which tells all the workers what jobs to do, the cell also has a "main office," called the *nucleus* (new-klee-us). In this small structure, often buried somewhere in the middle of the cell, are the "brains" of the factory. For the nucleus contains an amazing chemical called *deoxyribonucleic* (dee-ox-ee-rye-bo-new-CLAY-ic) *acid* or DNA for short. DNA is like a master set of blueprints, which holds the instructions for all the workings of the cell—all the chemicals that will be made and the reactions that will take place.

The nucleus is suspended in a sea of cell liquid, called *cytoplasm* (SYE-toe-plas-um). And in this sea there are many other structures as well. Some of them burn fuel and give the cell energy. Others manufacture chemicals.

There are still many unsolved mysteries about what a cell does. But scientists are constantly studying cells in their laboratories, and each day they are learning more about them.

The cells of our bodies work well together. We can walk and talk and play because the many different kinds of cells cooperate, each doing its share. But some cells in the living world live all alone. And they must do all their tasks by themselves.

4

It seems amazing that a tiny single cell, so small that it cannot be seen without a microscope, can do almost all the things that we can, with our trillions of cells, all working together. Cells that live alone can make or catch their food, and get rid of their wastes; some can move about; and they can even make new cells, just like themselves.

2
Cells That Live Alone

There are invisible worlds all about us. In each of these worlds, creatures so tiny that we cannot see them, move about, eat or are eaten, and give birth to new cells like themselves.

These busy little worlds can be found in a bucket of rainwater which has been standing a few days, or in a puddle by a roadside, a day or two after a rainfall. Countless numbers of tiny creatures swarm in the soil that we walk upon each day. Other creatures ride through the air in tough little "sleeping bags," called *cysts* (SISTS). If they happen to fall into a little puddle of water or even a moist patch, they spring into life. Tiny creatures are also to be found swarming in our own bodies.

What are these invisible creatures? With the help of a microscope we can find out what they look like and how they live.

Let's take a single drop of water from a pond or a puddle, and peer inside. Through the microscope we can see a whole new world of life.

Slipperlike creatures glide through the water, with a thousand tiny threadlike "paddles" called *cilia* (SILL-ee-uh) swishing to and fro in beautiful rhythm. These are *paramecia* (par-a-MEE-see-uh). In the same drop of water there may be other creatures, round or oval or pear-shaped, all moving about with the same kind of cilia.

But in the same drop of water, we may also find some very different kinds of creatures—little blobs without shapes, or really with a thousand different shapes. These are *amebas* (a-MEE-buhs), and as they ooze along, they change their shapes constantly. A little bulge appears in the ameba's jelly-like body, and grows, until the creature seems to flow into it. These bulges are called *pseudopods* (SUE-do-pods), which means "false feet." Sometimes they are like legs, on which the ameba crawls along. Other times they are like arms, with which the ameba catches its prey—smaller water creatures.

As we peer through the microscope, a little green-spotted *euglena* (yew-GLEE-nuh) may dart into view, swishing through the water with a tail like a lashing whip. This tail is called a *flagellum* (fla-JELL-um). The euglena does not have to catch

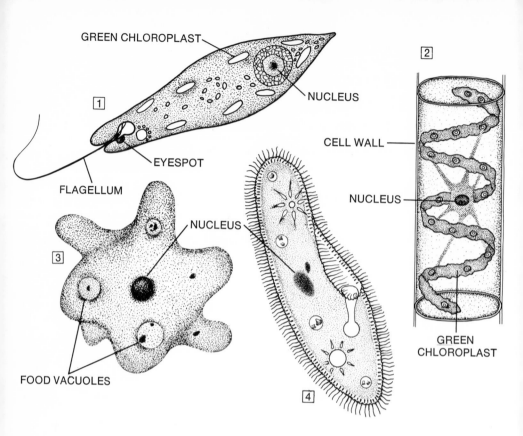

GREEN CHLOROPLAST

NUCLEUS

1

EYESPOT

FLAGELLUM

2

CELL WALL

NUCLEUS

NUCLEUS

3

FOOD VACUOLES

GREEN
CHLOROPLAST

4

Single-celled organisms: 1-Euglena; 2-Spirogyra; 3-Ameba; 4-Paramecium.

its food. It can make its own. All it needs is sunlight, and a few simple chemicals.

Some other creatures in our drop of water can also make their own food if they have sunlight. But they cannot move at all by themselves. These creatures, called *algae* (AL-gee), just float along, some by themselves and some in little groups.

8

All these microscopic creatures are single cells, but each can get along quite well by itself. Though they have no eyes, they seem to know just where the light is. Though they have no ears or nose, many seem to know just when a friend or enemy is near. Without any real arms or legs or fins they can swim after a choice bit of food and gobble it down. And when they grow large they can split in half; where there was one there are now two perfect little copies of the first.

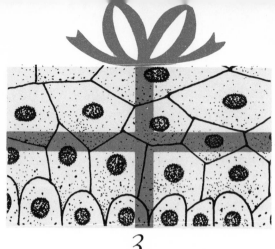

3
Cells That Cover

Your whole body is filled with cells. Trillions and trillions of these cells are busily working day and night to keep you alive. Your heart and your lungs and your liver are made of cells. So are all the other organs in your body. Each organ is wrapped in a covering of special cells called *epithelium* (ep-i-THEEL-ee-um).

In fact, your whole body is wrapped in a covering of epithelial cells, which is called *epidermis* (ep-i-DER-mis). (The epidermis is the outer part of the skin.) Some epithelial cells are flat like pavement stones. Others are like little cubes or columns.

Not only is the outside of your body covered with epithelium, but the inside is also lined with epithelial cells. The inside of the blood vessels is very smooth. The linings of epithelium let the blood flow easily and keep the delicate red blood cells from being harmed. Inside your mouth and nose, and in your lungs and stomach there is epithelium, too.

Every minute of the day you lose many of your cells. Every time you wash your hands, thousands and even millions of epithelial cells are washed away, down the drain, never to be seen again. Each time your sock rubs against your foot, you lose some cells.

"Why doesn't it hurt?" you may ask. The cells that you lose are already dead. All the skin that you can see is made up of dead epithelial cells. This is a lucky thing, too, for lying on this skin are many dangerous germs. If they could, they would slip inside your body and make you ill. But the dead cells of your skin are like a solid wall that keeps these harmful germs from getting through. (If you should cut yourself, you are making a hole in this wall, and germs may slip inside. That is why a cut can become infected.)

Our skin does much more than just cover and protect us. Some of it contains tiny oil glands. These produce an oily substance that helps keep the skin smooth. Without oil glands, our skin would dry out and flake away, and we would lose skin cells much more quickly than we do.

The skin also contains other glands, called sweat glands. It is through these glands that we perspire.

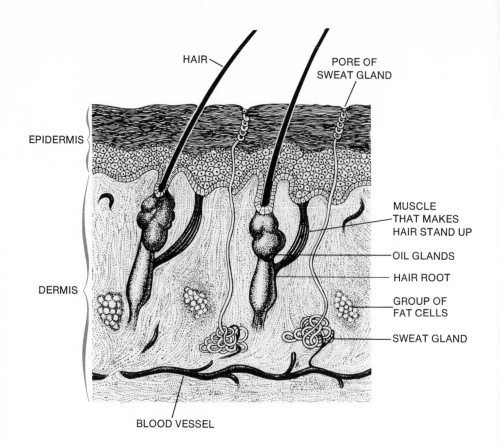

HAIR

PORE OF
SWEAT GLAND

EPIDERMIS

MUSCLE
THAT MAKES
HAIR STAND UP

OIL GLANDS

HAIR ROOT

DERMIS

GROUP OF
FAT CELLS

SWEAT GLAND

BLOOD VESSEL

Cross section of human skin.

Each sweat gland sends water out of the body through tiny openings in the skin called pores. Perspiring is very useful to the body in several ways.

You must have noticed that you perspire much more when it is hot. The sweat glands are an important part of the body's fine system for temperature control. (This system works so well that, except when you are sick, your body temperature

12

is usually 98.6°.) When it is too hot outside, the sweat glands work hard and send out more and more water. This water lies on the surface of the skin. Gradually it evaporates—it turns from a liquid into a gas—and goes off into the air. But evaporation is a process that uses up heat energy. The heat energy comes from the body. And so the evaporation of the water sent out by the sweat glands takes heat away from the body and thus cools it.

You know that during hot weather, when your sweat glands are working especially hard, you often get very thirsty. You must drink more water than usual, to make up for the water you are losing by perspiring. But did you know that you should also eat more salt than usual? For the water that people perspire is not pure water. It also contains a number of salts and other chemicals. This is another way in which the sweat glands help to keep the body working properly. For some of these chemicals are waste products, which the skin thus helps to get rid of.

In addition to oil and sweat glands, skin may contain the roots of hairs, which grow out from the surface of much of the body. On some parts of our bodies, such as the tops and backs of our heads, the hair grows very long. But on other parts, the hairs are very fine and very short. Although these hairs do not seem to be of much use to us, some of them have very important jobs to do. Thick, stiff hairs that grow out from the edges of our eyelids help to protect our eyes. Short hairs that grow inside our noses help to screen out dust and other small particles that we might otherwise breathe into our lungs.

Skin cells may also contain tiny bundles of a dark-colored chemical or pigment called *melanin* (MELL-a-nin). This pigment helps to shield us from the strong rays of the sun. When we are out in the sun, the bundles of melanin in the skin get larger, and our skin gets darker. That is what makes a suntan.

With all its different parts and jobs, the epithelium that covers our bodies is a very important part of us. And the epithelium that lines the inside of our bodies and organs is very important too.

Without the epithelial cells lining our intestines we would starve, even if our stomachs were filled with food. These special cells let tiny bits of digested food pass through and into our blood. There they are carried to the proper places in our bodies where they may be used for energy or to help us grow.

Another important kind of epithelium in our bodies is that which lines the many tiny air sacs in our lungs, which are called *alveoli* (al-VEE-oh-lye).

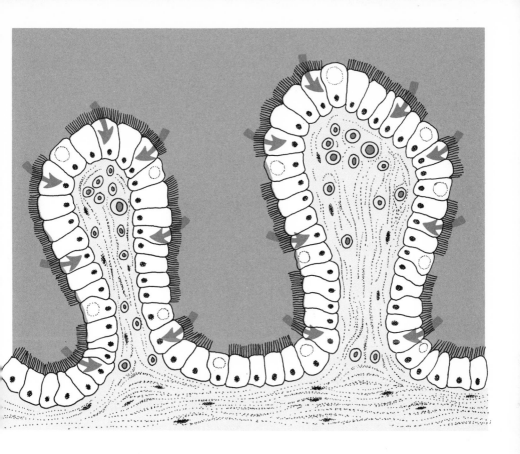

The cells that line the intestines help in the absorption of food.

The epithelium lining the alveoli is so thin that gases can pass freely through it. That is how the oxygen in the air we breathe into our lungs gets into our bodies.

Nearly all other animals have epithelial cells covering and lining their bodies too. In many cases, these cells are very much like ours, and they do the same sort of jobs.

Animal skins often contain melanin, and sometimes other pigments as well, which give their bodies different colors. Some animals, such as chameleons, flounders, and frogs, can even change their color, growing darker or lighter depending on how the tiny bundles of pigment are spread out in the skin cells.

Hairs grow out of the skin of many animals. Sometimes these hairs grow long and thick, giving the animal a coat of fur or wool that helps to protect it from the cold. In other types of animals, skin cells make feathers or scales, or even shells as hard as stone.

One of the strangest outer coverings in the animal world is that of the insects. Their skin cells make a tough outer cover containing a substance called *chitin* (KITE-in). This chitin coats an insect from head to foot, and even the inside of its digestive system as well. The chitin coat is like a suit of armor which helps to protect the insect. It has many joints, so that the insect can move about. Strangely, the chitin coat is also the insect's skeleton, for insects do not have bones as we do. It is their chitin "skeletons on the outside" that hold their bodies up.

Suits of armor provide good protection, but they have their disadvantages, too. They are very heavy to carry around. That is one reason insects are so

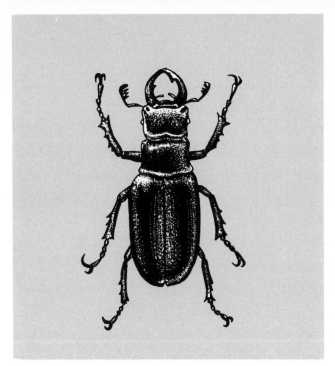

A beetle's body is covered with a jointed suit of armor made of chitin.

small. If their bodies grew much bigger than a few inches, their chitin coats would crush them.

Suits of armor do not provide any room to grow. And so, in order for an insect to grow, it must shed its skin, or molt. When an insect is getting ready to molt, it forms a new skin under its old one. Then its old chitin coat splits, and the insect wriggles out. The new skin underneath is still soft, for its chitin covering has not hardened yet. Quickly the insect breathes in all the air it can hold. It puffs its body up as much as it can, forming a larger framework for the new chitin coat to harden about. Within this larger shell there is more room, and new cells can grow to fill it.

Plants have covering cells too. The outside of a green leaf or stem is covered with a layer of cells called *epidermis*. The cells of this outer layer make a waxy substance, which acts as a smooth, water-proof "raincoat" for the leaf.

On the underside of a green leaf, there are special epidermis cells called *guard cells*. These come in pairs and look like two tiny kidney beans, placed side by side so that there is a small hole in the middle. There are thousands of them on the bottom of each leaf. The guard cells are like little mouths, opening and closing, according to how much sunlight there is and how much moisture is in the air. When the guard cells are open, gases can pass in and out of the leaf. And so, it is through these thousands of little openings that the plant breathes.

The roots of the plant are covered with a very delicate epidermis. These root cells act rather like the epithelial cells that line our intestines. They take in water and salts from the soil. Plants use the water and salts, along with gases that come in through their guard cells, to make their own food.

Thus we see that all living things are covered with cells on the outside and lined with cells on the inside.

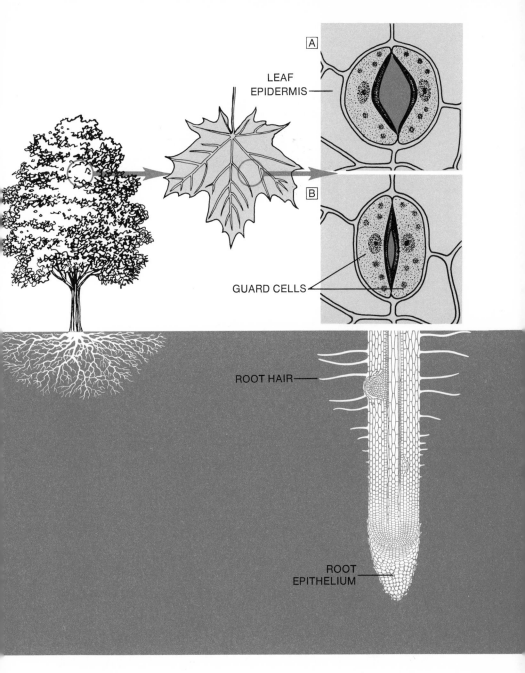

LEAF
EPIDERMIS

A

B

GUARD CELLS

ROOT HAIR

ROOT
EPITHELIUM

Guard cells in the leaves let gases in and out of the plant; water and salts are taken in through the cells covering the roots.

4
Cells That Travel

The many cells of the body need a steady stream of fresh supplies. They need building materials and energy. They must get rid of the wastes that build up.

For this the body has many highways—really waterways—the blood vessels. Every second the heart pumps fresh blood through the body. This blood flows along through branching blood vessels, which become smaller and smaller. Each cell of the body has a tiny blood-vessel waterway running right by it. It is here that the fresh supplies are delivered to the hard-working cells, and their waste products are unloaded into the blood stream.

The supplies are carried along in different ways. Sugars and salts float right along in the blood itself. So do other things that the cells need to help them grow. But the gas oxygen, which the cells need to

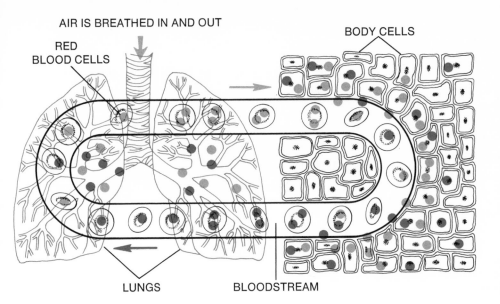

AIR IS BREATHED IN AND OUT

RED
BLOOD CELLS

BODY CELLS

LUNGS   BLOODSTREAM

The gas exchange. Red blood cells in the capillaries take up oxygen (yellow dots) from the air breathed into the lungs and give off carbon dioxide (gray dots) which they have collected from the cells of the body.

get energy, is picked up at the lungs by little "ferryboats"—the red blood cells. Their red color comes from a red chemical called *hemoglobin* he-mah-GLOH-bin). It is this chemical that can combine with oxygen and hold it inside the red blood cells.

The red ferryboats float along in the watery blood and deliver their cargo to nearly all the cells of the body. As the hemoglobin lets go of its oxygen, it is free to join with another gas. And so the red-cell ferryboats pick up a new load at each cell—the gas, carbon dioxide. This is a waste product made by the cells, and they must get rid of it, for if too much carbon dioxide builds up in a cell, the cell will die. The red cells carry the carbon dioxide back to the lungs, where we breathe it out into the air.

The red blood cells are not the only cells that travel in the blood stream. For every thousand red blood cells there are one or two white blood cells. These are very different from the red blood cells. The red cells look like little doughnuts, without the hole in the middle, but the white blood cells have no fixed shape at all. They look like a glob of jelly—just like the amebas that live in a pond. And like the amebas, they can swim or creep along, and

Two kinds of white blood cells.

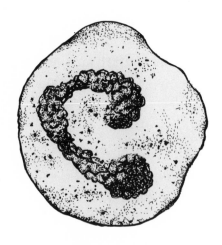

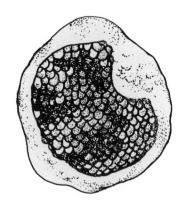

NEUTROPHIL   LYMPHOCYTE

can change their shape. They can even gobble up smaller living cells, just as amebas do.

What kind of cells do the white blood cells gobble up? They swallow up dangerous invaders, for every day our bodies are attacked again and again by bacteria, tiny cells much smaller than our white blood cells. Even though our skin helps to keep them out, these germs are always lurking about and slip in whenever they can—in the air we breathe, in the food we eat, or through cuts and scrapes.

These bacteria thrive on the warmth and food they find in our bodies. But they are unwelcome guests, for they multiply very quickly, and can swarm through our blood by the billions. Each one carries a bit of poison, which can make us ill and might even kill us.

But our white blood cells seek out bacteria. With some chemical sense, they seem to know just where these invaders are to be found. When a white cell meets a bacterium, it sends out pseudopods, just like an ameba catching its prey. It seems to flow around the bacterium, and then it swallows it up. The white cell seeks out more bacteria. Soon it has gobbled up five or six, perhaps even ten or more. The poisons in the bacteria that it has eaten finally kill the little warrior. But another white blood cell takes its place.

23

5
Cells That Help Us Move

Put your hand to your chest. Do you feel a thumping? That is your heart. It is really a muscle, beating away day and night, sending rich red blood through your body.

In addition to your heart, there are muscles in just about every part of your body—in your arms and your legs, in your neck and your back, and deep inside you. All these muscles are made up of thousands or even millions of muscle cells. These cells work together in bundles.

We could not move at all without muscles. It is the muscles attached to our bones that help us to move our arms or legs, to bend over or straighten up. Muscles also help to support us. Perhaps you have seen a skeleton. Unless it is carefully tied together and hung from something, it will collapse into a pile of bones. In a living human, it is the muscles pulling on the bones of the skeleton that keep the body standing up.

When we want to move our legs or arms, the muscles in them squeeze together. We say that these muscles contract.

When muscles contract, they help us to do work. We can lift things with our arm muscles and chew our food with our jaw muscles. Muscles move our eyeballs and thus help us to see. There are even muscles in our stomachs, which help churn the food.

All cells can contract a little, but muscle cells are especially good at contracting. They are so good at it that a man can lift a heavy weight. He can lift things that weigh more than he does.

Many muscles come in pairs. In our arms there is a muscle called the *biceps* (BY-seps). This is the muscle you see bulging out when a strong man bends his arm. While the biceps helps us to bend our arms, we could not straighten them out again if it were not for another arm muscle, the *triceps* (TRY-seps). The biceps and triceps take turns working. When the biceps is contracting, the triceps takes a rest, or relaxes, and our arm bends up. When the triceps is working, the biceps relaxes, and our arm straightens out.

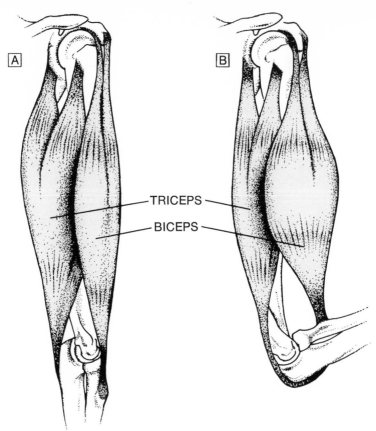

A. When the triceps contracts and the biceps relaxes, the arm straightens out. B. When the biceps contracts and the triceps relaxes, the arm bends up.

There are many different pairs of muscles in our bodies. Most other animals also have muscles. They are made up of bundles of muscle cells very much like ours. But in some animals the muscles are arranged quite differently.

In the earthworm, for instance there are two sets of muscles. One set is called the longitudinal muscles, because they run *along* the worm's body, from head to rear. The other muscles are circular muscles, which run in circles around the worm's body. The earthworm uses both sets of muscles to move along in its burrow in the soil. First its circular muscles contract. This makes the worm's body long and thin, and it stretches forward. Then its longitudinal muscles contract, starting from its head and moving back toward the rear. In each place where the longitudinal muscles are contracting, the earthworm's body becomes shorter and bulges out in a fat ring. Now its body is so fat that it can easily grip the sides of its burrow. And so, when the contractions of the longitudinal muscles have made its body shorter, it has moved up in its burrow. Then its circular muscles contract again, it stretches out ahead, and the whole process starts over again.

Other animals have different arrangements of muscles, depending on the types of bodies they have. A clam has special muscles that hold its two shells together. These muscles are so strong that a man cannot pull the shells of a live clam apart without actually cutting the muscles with a knife. The scallop, a relative of the clam, can also hold its shells tightly together. But it also has muscles that permit it to snap its shells quickly open and closed. Little jets of water are sent out when the scallop snaps its shells shut. This rapid opening and closing of the shells enables it to go scooting along in the ocean. (The clam cannot swim. It can only move very slowly along on the bottom, by rippling contractions of a large muscle in the soft part of its body, called the *foot*.)

The way an insect's muscles are arranged seems very strange to us humans. Our skeletons are deep inside our bodies, and our muscles are attached to the outer surfaces of our bones. But an insect does not have bones; its skeleton is the tough chitin coat on the outside of its body. Therefore its muscles must be attached to the inside of its skeleton, that is, to the underside of the chitin coat. Though this arrangement is very different from ours, it works very well. Many insects can fly, and it is muscles that make their wings move. Other insects can jump amazing distances in comparison with the lengths of their bodies. A man can jump about twice his own body length, and a kangaroo can jump five times the length of its body. But a grass-hopper can jump twenty times the length of its body, and a tiny flea can jump 200 times its own body length! If a man could jump as well as a flea, he could cover more than two city blocks in a single leap. It is the contractions of the muscles in the hind legs of fleas and grasshoppers that power their amazing jumps.

Although nearly all animals have muscle cells, plants do not have any at all. Plants move only by growing. And this is a far slower movement than that which muscles make possible.

The muscles of our bodies, and those of other animals could not work at all if they did not receive messages telling them what to do. Let us find out more about where these messages come from and how they travel.

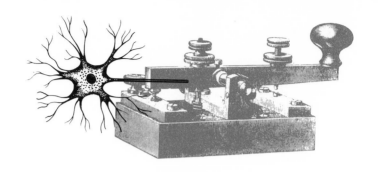

6
Cells That Carry Messages

There are miles and miles of "wires" in our bodies. Like the wires of a telephone company crisscrossing through the city, the "wires" in our bodies carry messages to and fro. These "wires" are not really wires at all, but long thin cells, which are called nerve cells, or *neurons* (NURE-ons).

Although some nerve cells may be more than a foot long, we cannot see them without a microscope. How can this be? Nerve cells are very thin —thinner than the thread you sew with; thinner than the hairs on your head; thinner even than the strands of a spider web. Some neurons are very short, only as long as the period at the end of this sentence. But others are longer than your arm. Some nerve cells in giraffes are more than eight feet long.

Just like telephone wires, our nerve cells are covered with layers of insulation, which keep the messages from going astray. These layers are made of a fatty substance and are wrapped around the neurons like a sheath. Scientists call this the *myelin* (MY-uh-lin) sheath.

But the nerve cells are different from wires in many ways. They can carry messages in only one direction. And since they are alive, they must be fed all the time.

There is a whole network of nerve cells in the body. The centers of this network are the brain and a long cord of nerves that runs down the back, called the *spinal cord*.

Some nerve cells carry messages *to* the brain and spinal cord from all parts of the body. Many of these carry messages from the eyes and ears and nose—the sense organs. That is why these nerves are called *sensory neurons*.

Other nerve cells carry messages *from* the brain and spinal cord to other parts of the body. These messages are often signals that tell glands to produce chemicals and muscles to contract. Without these messages we could not move. And the nerves that carry them are called *motor neurons*.

When we see a bird flying overhead or a car whizzing by, our eyes send messages to our brains

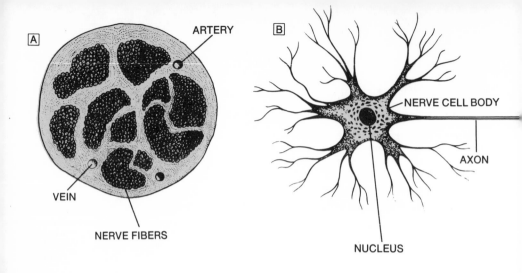

A. Cross section of nerves. B. The nerve cell.

in a certain code. Our brains then decode these messages, and we see. Whenever we want to move a foot, a hand, or even a finger, the brain sends a message to just the right muscles, and we move.

Sometimes we do things without thinking about them. If you touch a hot stove with your hand, you will pull your hand back before you know it. Only afterwards will you feel the pain from the heat. The message first went to the spinal cord. And a second or so later—after you had moved your hand—your brain found out about the hot stove. We call this kind of action a *reflex action*. The spinal cord is in charge of many of our reflex actions.

We could not live very long without our nerves, for the messages that they carry help us to breathe,

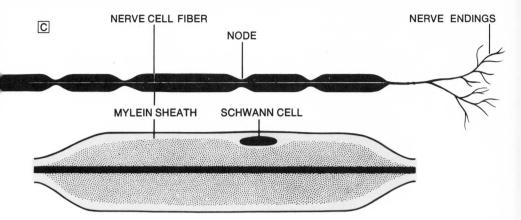

C. Close-up of part of the nerve fiber, showing the sheath that covers it like the insulation around an electric wire.

keep our hearts beating, and help keep our body parts working together as a team.

Our brains are filled with billions of nerve cells. They help us to think and remember, to love and to hate, to be happy or sad, to be angry or frightened. It is really our brains and what goes on inside them that makes us different from one another. Many people think that the most important cells of our bodies are the nerve cells.

Scientists do not yet know exactly how our brain cells work. But they are studying the brains of men and other animals, and each day they are learning more. Some scientists think that a special chemical called *ribonucleic* (rye-bo-new-CLAY-ic) *acid* or RNA somehow helps us to think and remember things. They are now trying to find other chemicals that will help our bodies to make more RNA. If they succeed, this may be a way to make us smarter.

Since nerve cells are so important to us, it seems surprising that plants have no nerve cells at all. Yet they are alive—they move and breathe and grow. The messages in plants are carried by chemicals. And they travel much more slowly than the messages carried by our nerves. That is one of the reasons why plants move more slowly than animals. For just about all the animals have nerve cells except the one-celled creatures.

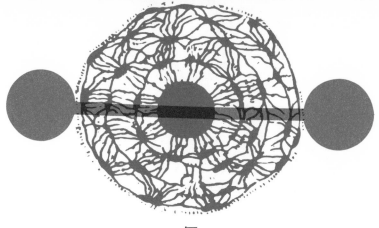

7
Cells That Build Strength

Have you ever watched a house being built? The framework goes up first. Your body also has a framework that helps to hold you up. Your framework is made up of bones, and it is called a *skeleton*.

Look at the picture of the human skeleton. The bones are of many different sizes and shapes, but they all fit so well together.

A house needs a firm "skeleton" that will hold up its walls against wind and storm. Of course a house does not move. But we do. Our skeleton must not only support us, but it must allow us to move as well. Many of our bones are put together in what we call *joints*. These joints let them move back and forth or around and about.

35

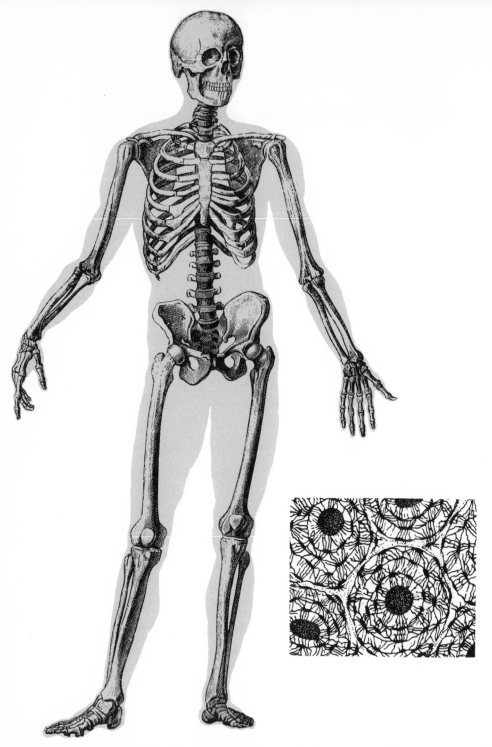

View of the human skeleton with a close-up cross section of a bit of bone.

If you looked at a bone, you would first think it was solid as a rock. But if you were able to look inside, you would have a big surprise, for the inside is hollow, and it is filled with many red and yellow cells. These cells make up the *bone marrow*. The spongy bone marrow is very important to us. The red bone marrow cells make red cells for our blood, and the yellow bone marrow cells make the white blood cells.

The rest of the bone is not solid either. If we could look at it through a magnifying glass, we would see many tiny circles. If we placed a tiny piece of bone under a microscope, we would see an even smaller circle inside each of them. And inside this smaller circle there are still smaller and smaller circles. They look something like the circles you see in a slice of onion.

Here and there we see tiny cells. These are the bone cells, which help build the bone. They manufacture a liquid that hardens like cement. Suspended in this cement-like substance are crystals of a salt, composed of the elements calcium, phosphorus, and oxygen. This hard substance is the bone.

Some bone cells make new bone cells. That is how your bones grow. And if you happen to fall and break a bone, these bone-building cells will make new cells to heal the break.

Other bone cells do a strange thing. They actually *eat* bone and help you grow. For you to grow and stay healthy, you must have plenty of marrow. But if the hollows inside your bones remained just as small as when you were born, they would not hold enough marrow to make your blood cells when you got bigger. Your bone-eating cells take care of this: as your bones grow, they make the hollow larger and larger.

Many animals have skeletons very much like ours. Of course, these skeletons may be of different shapes, and contain different numbers and sizes of bones, according to the size and shape of the animal. But the bones of mammals, birds, reptiles, amphibians, and fish are all basically similar to human bones and are put together in much the same way.

Some animals, however, have quite different kinds of skeletons supporting their bodies. We have already learned that insects have chitin skeletons on the outside. The bodies of certain other types of animals, such as clams, snails, and crabs, are encased in shells made of lime or some other hard material. And some small, soft-bodied animals, such as the various worms, have no skeletons at all.

Plants do not have bones. But they do have a sort of skeleton that helps to hold them up. Trees

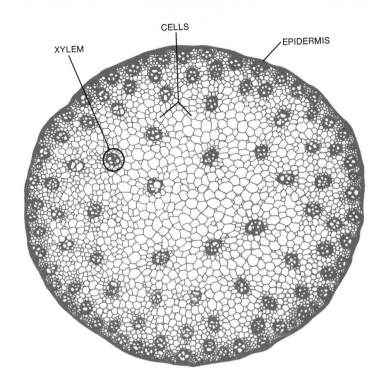

XYLEM   CELLS   EPIDERMIS

Cross section of a corn stalk, showing the different kinds of plant cells and the bundles of xylem vessels that help to support the plant.

are so firm that their wood is used to make furniture and even the frames of buildings.

The skeleton of a tree is made up of an inner core (which is actually most of the trunk itself, and a large portion of each branch) of *xylem* (ZYE-lum) tissue. The xylem in the outer layer of the trunk carries water and dissolved minerals up through the tree. But the xylem tissue in the core is dead, and its thick-walled cells are pressed tightly together to form solid wood.

The stalk of a daisy and even a blade of grass have skeletons too. Compared to a tree, the amount of xylem in these plants is very small. They have other types of tissue that help to stiffen them, but their skeletons are mainly made up of water!

This seems like a strange sort of skeleton indeed. But you can see how it works. Take a hot water bag and put a little bit of water inside. The bag is limp. Now fill it up all the way. It becomes hard and firm.

In a similar way, the cells of the plant take in enough water to make them firm. And it is this firmness of the cells which gives it strength. You can try another experiment to show how water helps to keep a plant firm. Cut off two plants at the base of the stem. Place one with its stem in a jar of water; place the other in an empty jar. In just a few hours you can see a startling difference. The plant in water still looks healthy. But the plant in the empty jar is drooping; its leaves and stem are limp. It has lost part of its water "skeleton."

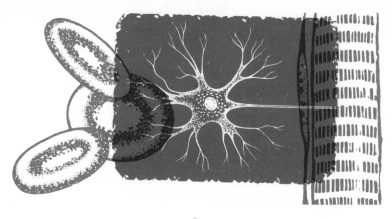

8
How Cells Work Together

The cells of your body are like the people of a nation. Each lives his own life, but each has a special job to do. While doing their jobs, the cells help one another. Just as a nation cannot get along without its doctors and truck drivers and carpenters, the body would not long live without its white blood cells, red blood cells, bone cells, and all the other kinds of cells that work together. The white blood cells fight disease bacteria; the red blood cells carry oxygen, carbon dioxide and other things; and our bone cells build bone frameworks for our bodies.

As each cell does its own job, it needs other cells to help keep it alive. All the cells must have food, get rid of their wastes, and be protected from injury and from attacks by foreign invaders.

41

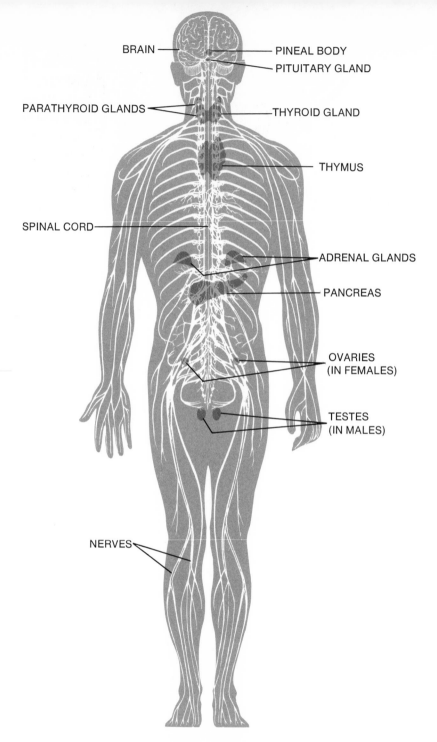

BRAIN

PINEAL BODY

PITUITARY GLAND

PARATHYROID GLANDS

THYROID GLAND

THYMUS

SPINAL CORD

ADRENAL GLANDS

PANCREAS

OVARIES
(IN FEMALES)

TESTES
(IN MALES)

NERVES

Hormones from the glands and messages carried along the
nerves control the activities of the body and its cells.

A nation must also have a government to keep everything working smoothly. The body too has a government, the brain. This is the central government of the body, which tells the cells when to do their jobs. The brain is helped to govern the body by the spinal cord and by different glands that are found in our bodies. We have already talked about how the spinal cord is in charge of many reflex actions. But the glands direct many different tasks in the body. They tell the cells how fast to burn food, and whether to store food or give up their supplies for other cells to use. They tell the heart how fast to beat and make the blood vessels widen or narrow. Glands help decide whether you will grow up to look like a man or a woman. They determine how fast you will grow. There is even a master gland that tells the other glands what to do.

The leaders and directors of the body must send their messages to the cells. For just as a nation has communication networks of radio and television and telephones and newspapers, the body has its communication networks too. Messages are carried from the brain and spinal cord along the telephone wires of the body, the nerves. And messages from the glands are carried in chemicals called hormones along the highways of the body, the bloodstream.

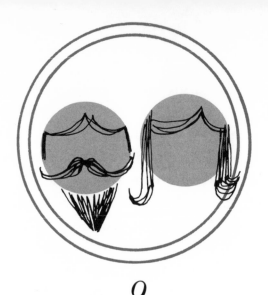

9
How Two Cells Become One

There has been life in this world of ours for billions of years, yet no single living creature has lived for more than a tiny fraction of this time. Many insects live for only a few weeks. A mouse rarely lives much beyond the age of two. Cats and dogs are already growing old at ten. Even men rarely live beyond a hundred years.

The history of life on Earth has been made up of long series of generations. Offspring follow parents, and then become parents themselves, in an endless succession. And even though no single creature lives for very long, its kind lives on in the new generations.

The long chains of life are continued by a process called *reproduction*. Indeed, this is one of

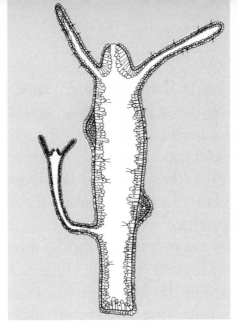

A hydra can reproduce by growing buds, small copies of the parent animal.

the most important characteristics of living things, which sets them apart from the nonliving. Animals and plants have many different ways of reproducing—of making more of their own kind. Tiny one-celled creatures, like amebas, paramecia, and euglenas, can reproduce simply by dividing in half. The one-cell becomes two "daughter" cells, and each one looks exactly like its parent. Small water animals called *hydras* (HYE-druhs) reproduce by budding. A tiny bud grows out from the parent animal and becomes a small, perfectly formed copy of the parent, which soon breaks away to live a life of its own. In budding, too, there is only one parent, and the offspring is exactly like it.

In most kinds of animals and plants, there must be two parents for reproduction to take place—a father and a mother. This kind of reproduction is called sexual reproduction, because it involves two different sexes, the male and the female. All the flowering plants and just about all the animals you can see reproduce sexually.

In sexual reproduction, the offspring looks a little like each of its parents. There is a good reason for this. For the life of a new offspring begins when two cells—one from each parent—join to become one.

The two cells that join to start the life of a new organism are very special cells. They are called *sex cells*, and they are quite different from the other cells of the body. The sex cell from the mother is called the *egg*, and that from the father is called the *sperm*. Eggs and sperms vary depending on the kind of organism. But generally they are very different from each other, in both size and shape.

Sperms are usually very tiny cells, which can be seen only with a microscope. It is their task to seek out the egg. And they are well-equipped for this task. For most sperms have one or more long, lashing tails, called *flagella*. (fla-JEL-uh). (These are the same kind of flagella that euglenas have.) A sperm with a long flagellum looks like a minia-

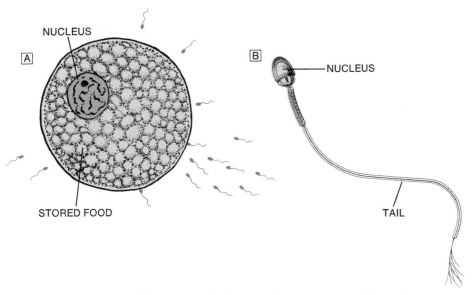

NUCLEUS

A

STORED FOOD

B

NUCLEUS

TAIL

Human egg cells are much larger than sperm cells. Only one sperm cell will win the race to fertilize the egg. A. Egg being fertilized. B. Magnified view of sperm.

ture tadpole. And like a tadpole, it can swim about quickly.

Eggs are usually larger than sperms—often thousands of times larger or even more. Eggs do not have flagella. They are just round balls, which cannot move about on their own. They are so much larger than the sperms because they contain stores of food, which can be used by the new offspring as it grows. Human eggs are about the size of a tiny speck—just barely large enough to see without a microscope. But the eggs of some animals, such as chickens, are much larger, for they contain much more stored food, which is called *yolk*.

The egg and the sperm both contain the chemical DNA. Each holds a complete set of instructions for a new organism. Although the egg is much larger than the sperm, each contains exactly the same amount of DNA. And so, when egg and sperm join, the new organism will receive one-half of the "plans of life" from its mother, and the other half from its father.

The joining of sperm and egg is called *fertilization*. This is an exciting process. First there has been a race, with thousands or even millions of sperms swimming toward the egg. But even the first sperms to reach the eggs are not the winners. For the egg has a tough coat around it, and the sperms cannot get through until the coat is broken. Each sperm carries with it a tiny "bomb" of a special chemical that helps to wear a hole in the coat around the egg. When enough sperms have released their tiny bombs, a hole is broken through the coat. Then the next sperm that reaches the egg will be the winner of the race, for that sperm will not stop at the outside of the egg. It goes right into the egg. Its DNA mingles with the DNA of the egg, and the life of a new organism is on its way.

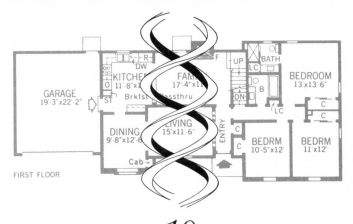

10
How One Cell Becomes Two

Everybody knows that if you plant a seed it may grow first into a small plant, and then, growing a little larger each day, it may eventually grow into a large tree. Every animal baby, too, is born small and little by little, it grows big like its parents.

It may seem strange, but the cells in your mother and father are just as small as the cells in your own body. Why then are your parents so much bigger than you? Their bodies are made up of many more cells than yours.

All living things, large and small, start life as a single cell. But, how does a single cell grow into a tree or a lion or a whale or a mouse, or you? Most living cells are able to split into two parts. Each part now becomes a new cell, smaller than its "mother." But soon it grows, and it too can divide in two.

Each tiny living cell contains DNA, the chemical that holds the set of blueprints, or plan for what the cell will look like and do.

When a tiny cell that some day will be a man starts its life, it has all the plans for the body: the shape, the size, the number of hands, and the number of fingers on each hand—everything about the man-to-be. This plan or code of life even carries instructions for each of the cells—the eye cells, the heart cells, the lung cells—all the different kinds of cells that work together in our bodies.

When a cell is getting ready to divide, it makes a perfect copy of its DNA, containing all its instructions. Then each "daughter" cell gets one of the copies. So we can see now that every cell in the body has exactly the same blueprints. If all the plans are the same, why do different cells of the body do different things? Why do the eyes see and the heart beat? It seems that each cell of the body can read only part of the plans—the part that tells it what its own special jobs are.

We all begin life as a single cell, called a *zygote* (ZYE-gote). This cell divides into two. Then each divides again, and we have four cells. The four soon become eight, then 16, 32, 64—before we know it, there are hundreds and even thousands of cells. At first the cells are all alike, and cluster together

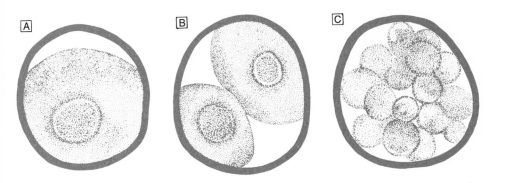

The zygote divides again and again to form a group of cells that will someday develop into a human being.

to form a simple ball. But as more and more cells are formed, differences begin to emerge. An embryo of a human being begins to take shape.

Soon a head appears, and then eyes. Before long, tiny buds appear, which will form arms and legs and even a tail. More and more cells are reading their own special parts of the blueprint. And soon a baby is born. It will grow and grow, as its cells keep on dividing. And at last it is a man. But even now, many of the cells in the body keep on dividing. For each day, billions of cells in our bodies die, and new ones must be born to take their place.

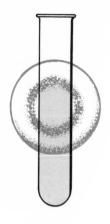

11
Cells in the Laboratory

Nearly all the secrets of life are locked up inside the cell. If we understood just how the cell works, we could solve many of the mysteries of how organisms live and grow, and why they die.

Thousands of scientists all over the world are studying the cell and its workings, both inside living things and in test tubes. Every day they are learning new facts—new parts of the great puzzle of life. Little by little, they are putting these pieces together and gaining a clearer and clearer picture.

Although cells live best inside a living organism, scientists have developed ways of growing them in test tubes quite well. These techniques are known as tissue culture. A group of cells of the same kind, working together, is known as a *tissue*.

Just as in the living body, a cell growing in a test tube has certain needs. If these needs are not met, it will die. The cell must be fed—it must have certain sugars and salts and other chemicals, and oxygen to help it burn its food for energy. Its waste products must be taken away—if they are allowed to build up, the cell will be poisoned.

If all its needs are met, a cell in a test tube will live and multiply, producing many more cells like itself. Some cells grown in test tubes can even go on to make a whole new organism if they are treated in a special way. Small bits of tissue from a carrot, for example, can be made to grow into a whole new carrot plant.

There are many advantages to working with cells in a test tube. Often the scientist wants to find out what effect a particular chemical has upon the life of a cell. In the test tube he knows exactly what chemicals he is adding, and he can study the effects of these chemicals upon the cell. But with a living organism, the effects of all the other chemicals that are already present in the cells might confuse him.

Scientists can also peer into the cell itself, using powerful electron microscopes. These are quite different from the microscopes that you may have seen, and they are much stronger. They use electrons instead of light rays, and objects seen through electron microscopes can be magnified to millions of times their real size. By taking photo-

graphs through these powerful instruments, scientists have gained valuable information about important structures inside the cell.

Many other instruments are also used by scientists to learn about cells. One of the most important of these is called the *ultracentrifuge* (ul-tra-SEN-tri-fewj). This instrument can spin things around at very high speeds. If you tied a rope to a bucket partly filled with water and then quickly swung the bucket in a circle around your head, you would find that (if you whirled the bucket around fast enough) none of the water would spill out, even though the bucket was turned on its side. A force pushing outward from the center of the circle would push the liquid out toward the bottom of the bucket and so keep it inside. This force, called *centrifugal* (sen-TRIF-yew-gal) force, is stronger, the faster the bucket is whirled around. You could probably swing a bucket around your head no more than a hundred times a minute. But an ultracentrifuge can spin a tube about more than a *hundred thousand* times a minute, and so the force it can exert is enormous.

An ultracentrifuge can be used to separate the different structures and chemicals inside cells. First the tissue is ground up into a soupy mash. This breaks the outer walls or membranes of the cells, allowing the cell fluids to flow out. Then the

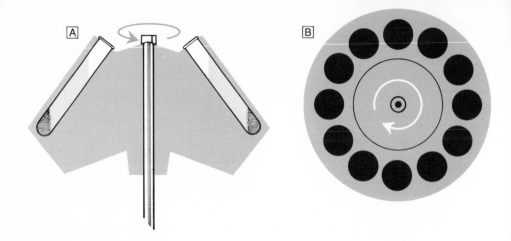

The centrifuge. A. Cutaway side view. B. Top view. The test tubes are spun around, and the heaviest parts of the mixture are deposited at the bottom.

mash is placed in a test tube and spun around in the ultracentrifuge at a high speed. The solid particles, which are mostly broken bits of the cell walls or cell membranes, sink to the bottom of the tube, and the liquid remains at the top. This liquid contains many structures and chemicals that were inside the cells. It is poured into another test tube and spun around again, at a higher speed. Now the structures and chemicals from within the cell itself begin to settle out. First the heaviest structures sink to the bottom of the tube. These can be separated from the liquid and studied. Then the liquid can be spun around again at an even higher speed, and the next heaviest structures or chemicals will come out of the liquid and settle to the bottom of the tube. The different structures and chemicals in the cell, each with its own weight, can thus be isolated by ultracentrifuging.

Using this and other techniques, scientists have been able to study many of the chemicals that are found in the living cell. Scientists have learned much from studying not only normal cells, but other cells that are not normal, such as cancer cells. In cancer, the cells run wild—they keep dividing and dividing, and the tissue does not stop growing, as normal tissue does. Scientists are not sure just why this is. But studies of cancer cells in the test tube are helping them to learn more.

Some scientists believe that with a complete understanding of the cell and how it works, man may learn not only the causes of cancer and other diseases, but how to control and cure them as well.

Index

fertilization, 48
flagellum, 7, 46, 47

glands, 43
guard cells, 18

hair, 13, 14, 16
hemoglobin, 21
hormones, 43

insects, 16–18, 28, 38

jumping, 28

lifespans, 44

melanin, 14, 16
motor neuron, 32
muscles, 24–29
muscles in
    clam, 27, 28
    earthworm, 26, 27
    insects, 28
    man, 24, 25
    scallop, 27, 28
myelin sheath, 31, 32

nerves, 30–34
neuron, 30
nucleus, 4

oil glands, 11
oxygen, 20, 21

paramecia, 7, 45
perspiration, 12, 13
plants, 18, 29, 34, 38–40
pseudopod, 7, 23

red blood cells, 22, 23
reflex action, 33
reproduction, 44–48
reproduction in
    hydra, 45
    man, 46–48
    one-celled organisms,
        45
ribonucleic acid, 33
RNA, 33

scallop, 27, 28
sensory neuron, 32
sex cells, 46–48
sexual reproduction, 46–
    48